The Commonweath ex rel. James Todd, vs. The Allegheny Bridge Co'y.	Quo Warranto. Brief for Defendant.

In this case the Jury found—

1st. That the defendants did, by special resolution, prohibit James Tod, the relator, and his family, from crossing and re-crossing said bridge at the yearly rates prescribed by the bye-laws of the corporation.

2d. That the defendants did not, every third year, make such returns as it was required to make.

3d. That the profits and income, after the decenneal period, did bear a dividend of more than 15 per cent., or *would have borne such dividend* if the rates of toll allowed by the charter had not been reduced.

4. That the defendants did vest the excess of clear profits over 15 per cent. in bank stock and other productive funds.

5. That the tolls over and above what defendants were authorized to receive and retain for the stockholders (and which might have been received) would not at a fair

appraisement have been sufficient to pay for and redeem the bridge, and pay the stockholders the appraised value of profits for the residue of the term of forty years.

6th. That the defendants did not appropriate the tolls to the payment of the individual indebtedness of the stockholders.

7. That some of violations set forth in the suggestion were committed within thirty years before the filing of the suggestion.

The quession now arises upon the whole record, what judgment is the court authorized to pronounce?

The record presents three demurrers by the plaintiff, and a verdict upon seven points.

The first demurrer is upon the ground that the company had no right to prescribe any rate of trip toll other than the rates specified in the act.

To this it is answered that the statute prescribed only a maximum rate, beyond which they could not lawfully exact. But it was incident to the general powers of the corporation to stipulate in good faith with parties for a different rate.

The second demurrer is to the plea setting forth the contract with General Robinson as to the use of his ground on which the abutments were erected. It is submitted that in this respect it was perfectly competent to the corporation to make such contract as might in the judgment of the directors be most advantageous to the carporation, and that the consideration received was a full equivalent for the privilege granted, being alleged in the plea and admitted by the demurrer, constitutes a full defence to that charge.

The third demurrer is to the plea stating the reasons for contracting a debt of 15,000 for the completion of the bridge.

To contract debts within the scope, and for the objects of its being, is incident to every corporation. In this instance the right to increase the capital stock was a privilege conferred—not the denial of a general right. From the peculiar circumstances of the country, as explained by Mr. Anderson, and averred in the plea that privilege was not available, and but for the exercise of its general right, the bridge, as appears, could not have been erected.

By contracting that debt the design of the State in creating the corporation, was fulfilled; the benefit of a valuable public work secured, while the company violated no duty and exceeded no powers by the charter granted.

Upon the demurerrs, then, the law is with the defendant. But were it otherwise, no ground of forfeiture remains unanswered; and the facts found by the jury establish a full and complete defence to every charge in the suggestion worthy of any consideration.

Regarding the finding of the Jury on the first three issues, submitted to their consideration, as supported and sustained by legal evidence, we deny that it justifies a forfeiture of charter or seizure of franchises.

The injury, if any, to the relator, charged in the first issue, is strictly private. No public right or interest is affected; no liability to forfeiture is incurred; no infraction of the charter is caused by the act charged.

The case of The People vs. the Hillsdale and Chatham Turnpike Company, 2 Johnson's Reports, 190, is, in our judgment, a conclusive authority against the Quo Warranto. Woodworth, Att'ney General, moved for a rule that the defendants show cause, by the next term, why an information, in the nature of a *Quo Warranto* should not be filed against them. He read affidavits stating that the road had been opened through the land of the complainants, and used, without any offer having been made to them to agree upon the compensation, and without having the damages ascertained according to law.

E. Williams, contra.

" *Per curiam:* If the defendants have not followed the directions of the act relative to the compensation to be made to the owners of land, through which the road had been made, they are trespassers, and the complainants have adequate remedy in the usual course of common law. The public are no way interested in the controversy or complaint, and that is a sufficient reason for not granting this extraordinary remedy. Rule refused."

We concede that a private and public remedy may co-exist. A private action and a public prosecution may in some cases both be sustained.

The same doctrine is laid down in the case of The People vs. Bristol and Renssalearville Turnpike Company, 23 Wendell's Reports 244-5. The principle now contended for (that there might be a remedy both by private action and public prosecution,) was argued in London vs. Vanacre, 12 Mod. 270-1. It was held, as we before noticed, that the the city must forfeit its francise of the Shieverick if it did not elect a Sheriff and compel him to serve. He had refused to serve. Holt said, as to the objection, that he may be indicted for this refusal, as was the case of Lanwood, Sh'ff of Norwich, I answer, that will not be sufficient to hinder the forfeiture of the franchise; for if there should be a vacancy when the Sheriff comes to be sworn, there will be an obstruction of justice. "*The ground was that a public inconvenience in not exercising the franchise would work a forfeiture, although there might be another remedy.*"

These cases clearly show the instances in which a private action and a public prosecution may both be sustained. This remedy is sought in the name and under the authority of the Commonwealth, and cannot upon the strength or reasoning of any authority adduced, be sustained upon allegation of any mere private injury.

We concede that "all franchises which are granted are upon condition that they shall be duly executed *according to the Charter, that being a condition annexed to the grant.*" This Company is not charged with the doing of any act prohibited by its charter; with any infraction of the law of its being.

"*The Charter is the law of the case.* If none of *its* provisions have been violated, it is difficult to find any legitimate ground to demand its surrender." *Corwin vs. Urbana Insurance Company*, 14 *Ohio Reports*, 10.

Again, we deny that there was any legal evidence before the jury justifying the finding of that issue for the Commonwealth. See the testimony on pages 1, 2, 5, (near the bottom) 10 and 11, relative to Todd, in Judge McClure's notes.

The testimony of Cook, page 11, McClure's notes shows conclusively that he did not act under any resolution of the Board excluding Todd and that he had never heard of any such resolution.

The testimony of Robert Hall relative to his conversation with Stoner, page 2, was wholly illegal.

Again it will be observed it is neither charged nor found by the Jury that Todd and his family were excluded from passing the bridge, paying the regular trip toll assessed under the Charter upon all trip passengers; but simply that he was not allowed to pass as a yearling. The commutation was not a right but a privilege of the act of the Board in the exercise of their discretionary power. No improper exercise of that power being charged, it is presumed that just cause existed. And so long as he was allowed to pass as required by the Charter paying the regular trip toll there was no right violated. Having kept themselves within the legal charge, the exercise of their discretion is not a subject for revision by the Court. But were it not so the evidence discloses a sufficient reason why Todd should have been required to pay by the trip.

The suggestion charged the exclusion of Tod *and his family;* there was not one particle of proof before the jury of any resolution or action of the board or any of its officers in relation to the family of Tod, so that the finding is without evidence to support it.

The second issue—if issue it may be called—is an immaterial one. The suggestion charged an obligation to make return once in ten years, and averred a neglect of the Company to make such return. The plea alleged that the Bridge was completed in 1819, and that, after the third year from the completion of the Bridge, the Company did from thence forward make returns each and every year. The replication avers that the defendants did not each and every third year make such return as they were required by law.

This is a total departure from the alleged cause of forfeiture charged in the suggestion; and if the truth of the finding of the jury be conceded, there is no cause of forfeiture involved, as the replication charges no violation of the corporate duty under the provisions of the charter. Conceding, then, verity to this finding, it is idle and wholly unavailing. No judgment of forfeiture or seizure could be legally or properly pronounced upon a finding involving no violation of charter.

Again. The finding is unsupported by any evidence whatever. The law presumed that the Company had performed its duty and made the requisite returns.

The Commonwealth proved, (page 1, Judge McClure's notes,) return from 1830 to 1850, except two years, 1844 and 1845.

The Defendants proved (page 7) by Harper, treasurer, that he made returns every year from 1842 to 1851; it is also stated on the Judge's notes, page 9, that Mr. Harper shows returns made to Auditor General's Office from 1822 to 1848, inclusive; and the defendants also gave in evidence, page 9, a return made by the Treasurer of the Bridge Company, in 1822, to the Legislature. This proof sustained literally every allegation in the plea, and went far beyond the averments of either suggestion or replication. Both legal presumption and proof were clearly with the defendants.

Again. We deny that a mere omission to make a return *in time* is a legal cause of forfeiture of corporate rights and corporate property.

The 14th section of the original act of Incorporation passed in 1810, required returns to be made every third year from the date of the act, until three years next after the bridge shall be completed; and if, at the end of three years after the completion of the bridge, it should appear that the average profits would not bear a dividend of six per cent., the rates of toll might be increased so as to bear that dividend. This section further provided that, at the end of every ten years, an abstract of the three preceding years should be presented; and if it appeared that the clear profits and income would bear a dividend of more than 15 per cent., the excess should compose a fund for the redemption of the Bridge, save a small toll or revenue to keep it in repair—the excess to be vested in Bank stock or other productive funds.

This act having expired, was revived by subsequent act in 1816.

The Stock was subscribed and the Bridge opened in 1819. Three years afterwards the first return was made. Thence forward an annual return was made down to the present time.

The object of the first return was to benefit the Company by permitting an increase of toll in a certain contingency.

The State, having become a subscriber, required by act annual returns to be made, so as to exert a more vigilant supervision.

This provision superceded that in the charter, or, if it did not, the annual returns actually made were a virtual compliance with the charter.

It is obvious, moreover, that this provision was merely directory; a duty imposed on the officers, for the faithful performance of which penalties were imposed on them. And had there been total neglect, it would not work a forfeiture of the charter. *Hillogg v. Union Co.* 12 Conn. 7.

Third Issue. Here again the replication goes entirely beyond the suggestion. The suggestion averred that the profits and income, after the first decennial period, bore a dividend of more than 15 per cent.

The plea denied the averment. The replication avers that the profits and income did bear such dividend, *or would have done, if the rate of toll allowed by the charter had not been reduced.* Here entirely new matter was introduced into the replication, and the Relator went before the jury upon a proposition not *at issue* between the parties.

The jury, however, found no violation of duty by the defendants—there was no illegality in the profits exceeding 15 per cent. at any time, especially here as the jury found on the next (fourth) issue, that the defendants vested the excess of clear profits over 15 per cent. in Bank Stock and other productive funds. That they might exceed that sum was contemplated and permitted by the act, inasmuch as it provided for the investment of the surplus.

The fourth, fifth and sixth issues, the only ones containing any serious charges against the defendants were found by the jury in their favor.

It is but sheer justice to the defendants to say that there was not one particle of testimony before the jury to sustain the gross and infamous imputations contained in the sixth issue, and that the same was contradicted by the testimony of Anderson and Snyder, and also by volnminous documentary evidence in the cause—demonstrating that the charges against this Company emenated from the mere grossness of malice.

The Relator expressly alleges in the first suggestion that the bye-laws in relation to yearly toll were enacted by the defendants "for their own government and for better carrying into effect the provisions of the Acts of Assembly" incorporating the Company.

This surely charges no assumption of power, neglect of duty, or violation of law, and in our judgment, requires no further argument.

Seventh Issue. To all the charges contained in he suggestion the defendants pleaded in bar of forfeiture that not any of them had been committed within twenty-one years next preceeding the filing of the suggestion.

That this plea, if true, is a full and complete bar to any judgment of forfeiture is well established by the following authorities: Angell and Ames on Corporations, 624. Sec. 5.

Now the truth of this plea is either confessed by the plaintiff, or what is equivalent thereto, it remains unanswered upon the record, and no judgment of forfeiture could be pronounced whatever might have been the charges alleged, and the finding of the Jury upon them. The relator instead of answering the plea, replied that some of the facts charged were committed within 30 years. By this replication he admits their non-commission within twenty-one years, and, having himself taken the record down for trial, leaving a good plea unanswered, he is thereby precluded from questioning its truth.

One further observation may be made. The State was one of the original Stockholders in this Company to the amount af $40,000, nearly one-half of the whole capital. It appears by the evidence upon the record, that in 1822, shortly after the Bridge was completed, a full statement was made to the General Assembly, through one of its Senators, Condy Reguet, of the condition and acts of the Company, its system of tolls and the reasons for its adoption.

From that time until 1843, a period of twenty years, an annual statement of the affairs of the Company was regularly made, and the State each year without complaint received its regular dividend. This acquiesence would seem to preclude all ground of forfeiture for any thing prior to that time. But in 1843 the State, by Act of Assembly, offered her stock in the market and sold to the present holders or those under whom they claim at a premium of about $8, on each share of $25. So that the receipts of the State from the forty thousand dollars entrusted to the management of this Company present the following result:

Annual dividends, - -	$70,000 00
State Tax, - - -	5,657 03
Proceeds of Stock Sale, - -	53,194 75
	$128,851 78

For the State to forfeit the Stock thus purchased from her by those in whose hands it now is, on the ground of antecedent irregularities however gross would be a violation of public morals and justice.

But the State in her own Courts is governed by the same legal principles that bind her citizens. Having bargained and sold this Stock for value she cannot avoid the sale and resume the franchises of the corporation for any alleged forfeiture prior to her sale.

Independent of this consideration it is claimed that no ground of reproach much less of forfeiture has occurred at any time before or since that sale; and that upon each article and upon the whole record, the defendants are entitled to Judgment.

A. W. LOOMIS, SHALER & STANTON,
Attorneys for Defendants.

www.ingramcontent.com/pod-product-compliance
Lightning Source LLC
LaVergne TN
LVHW020644110826
845149LV00004B/1350
* 9 7 8 1 4 1 8 1 8 9 6 2 4 *